Timmy Turtle's Lazy Eye

Timmy the turtle came out to play

Eye games with the doctor today

He tested one eye and then the other

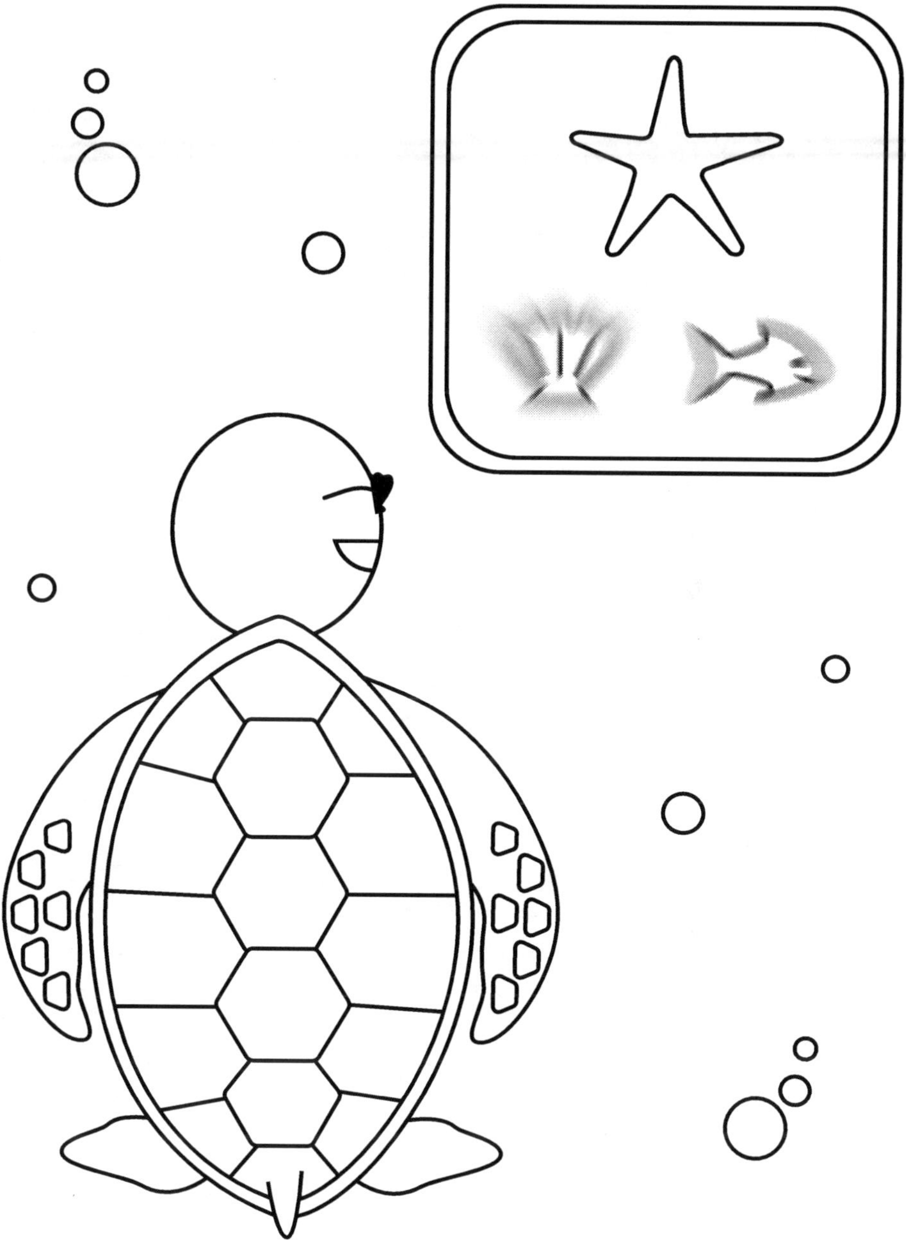

And found one was seeing fuzzier

He gave Timmy some eye patches to cover his better eye

So that his brain can make the fuzzy go bye bye

But Timmy has to use the eye patch everyday

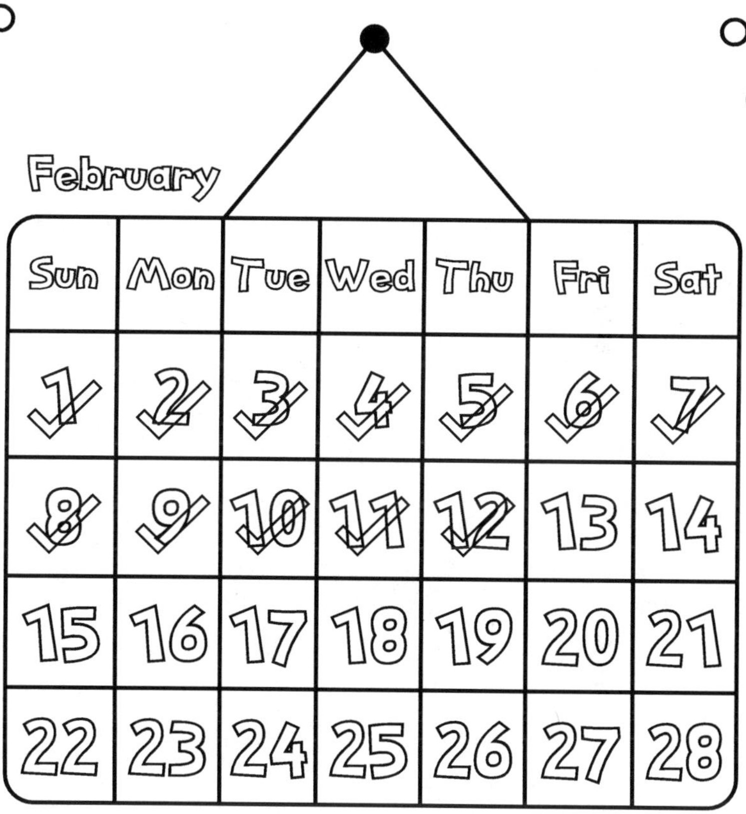

To make the fuzzy eventually go away

Sea School

He wears it when he's
at home and at school

As being a pirate
makes Timmy feel
quite cool

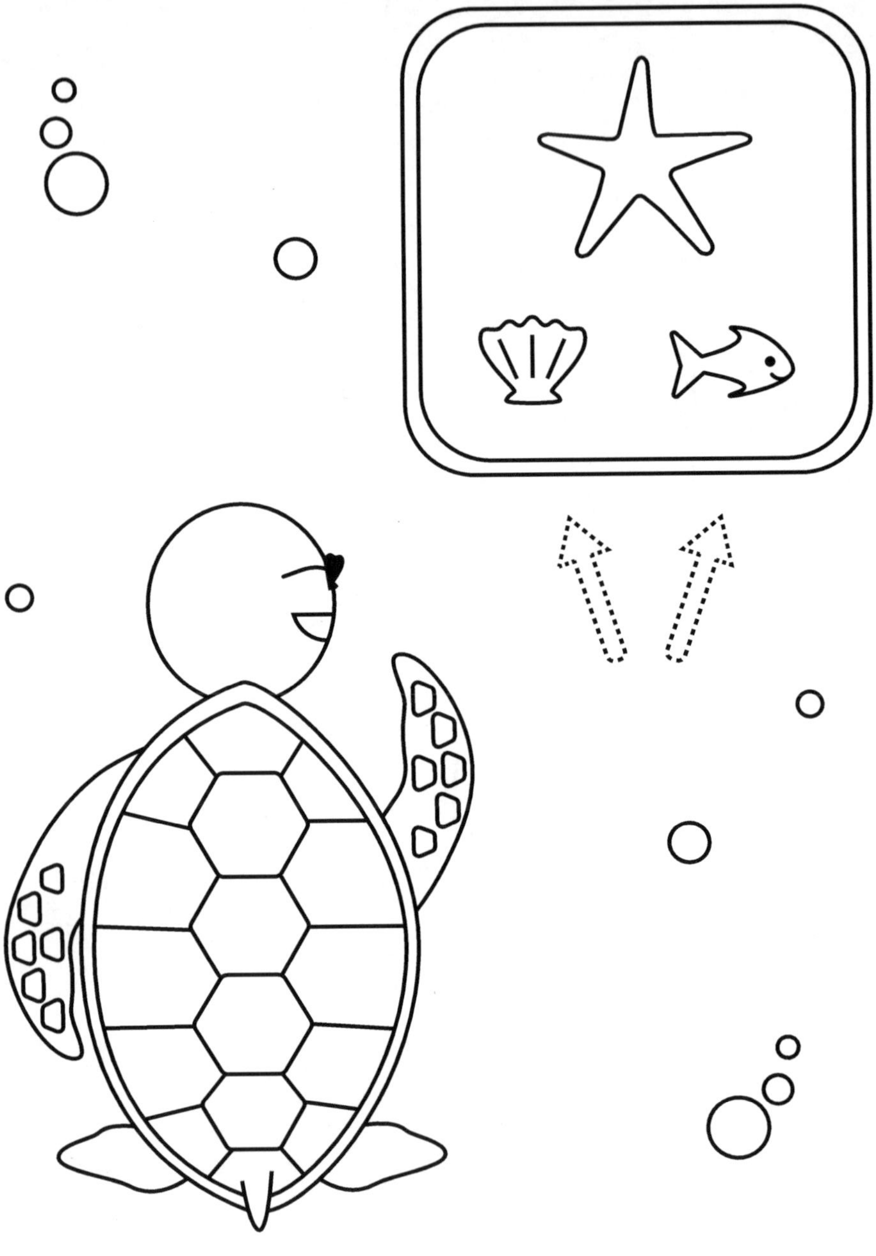

When Timmy visits the doctor next time

The doctor says, "Oh wow! Now you can see a smaller line!"

Timmy is happy that the fuzzy is going away

So he keeps using his eye patch everyday

Soon the doctor tells Timmy he doesn't need the eye patch anymore

And Timmy thanks him for making his eyes better than before

Section 2

Sealife Alphabet

In the sea from A to Z

But your doctor may use X O U Y H and V frequently

Aa

Anemone

Blowfish

Also known as pufferfish

Bb

Cc

Coral

Dd

Dolphin

Eel
Ee
Ff
Flying Fish

Gg

Giant Squid

Hh

Hermit Crab

Ii

Indian Starfish

Jj

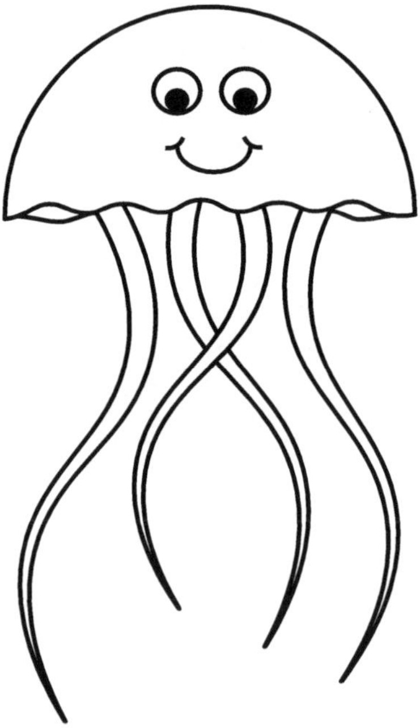

Jellyfish

Killer Whale

Also known as Orca

Kk

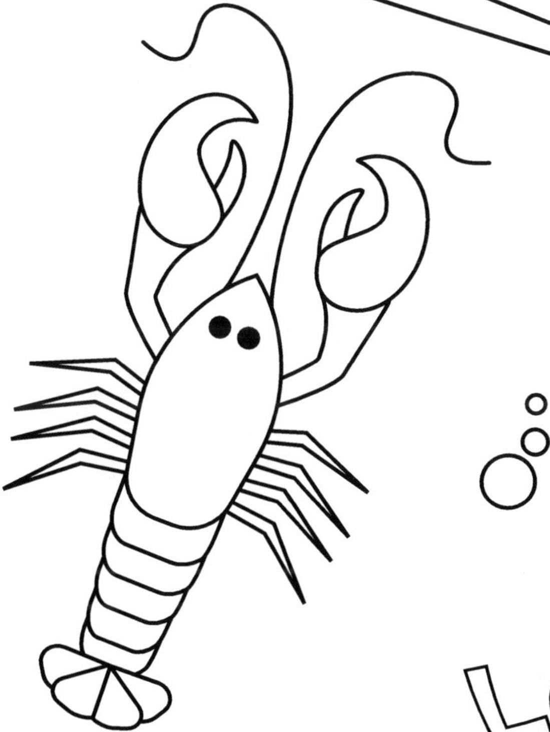

Ll

Lobster

Manatee

Mm

Nn

Narwhal

Oo

Octopus

Pp

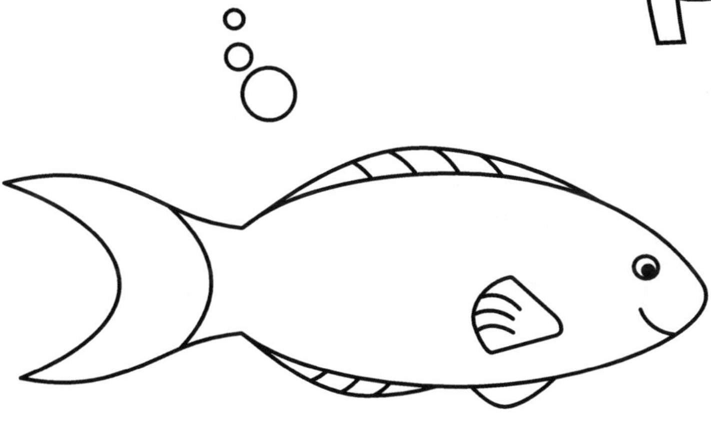

Parrotfish

Quahog

Also known as clam

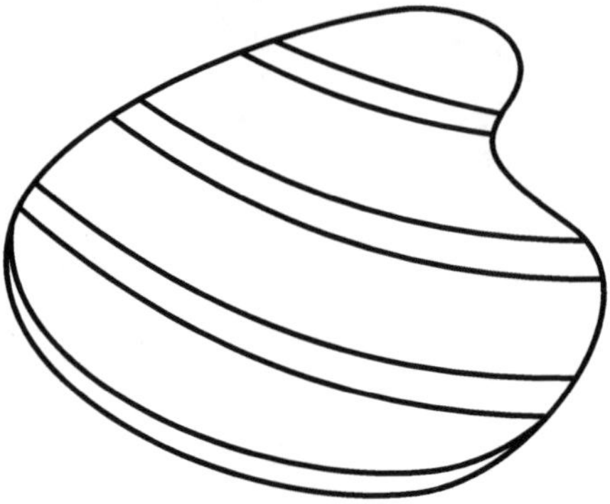

Qq

Rr

Razorfish

Ss

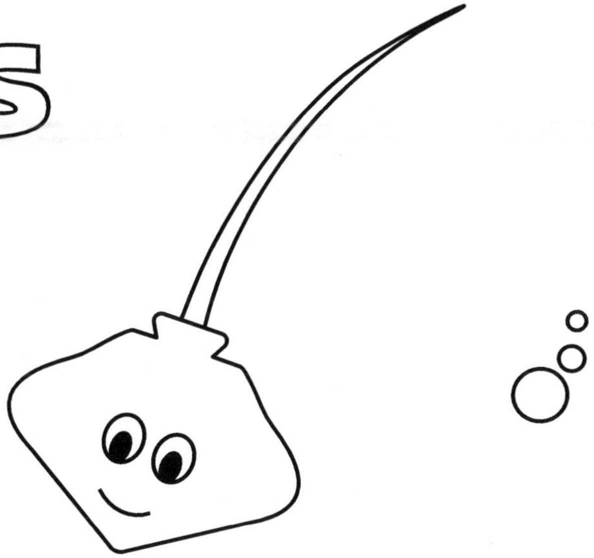

Stingray

Tt

Turtle

Uu

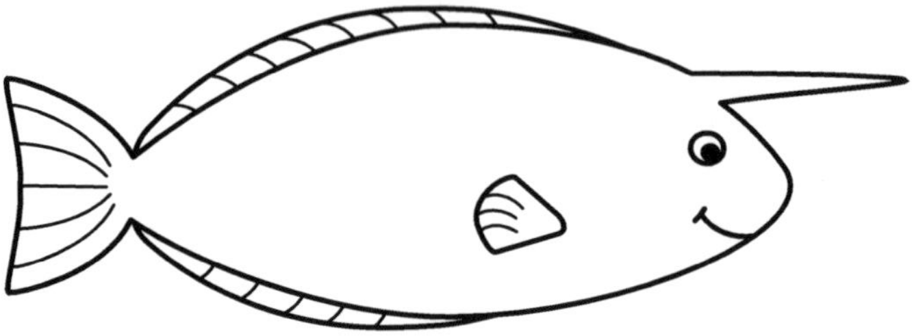

Unicornfish

Vv

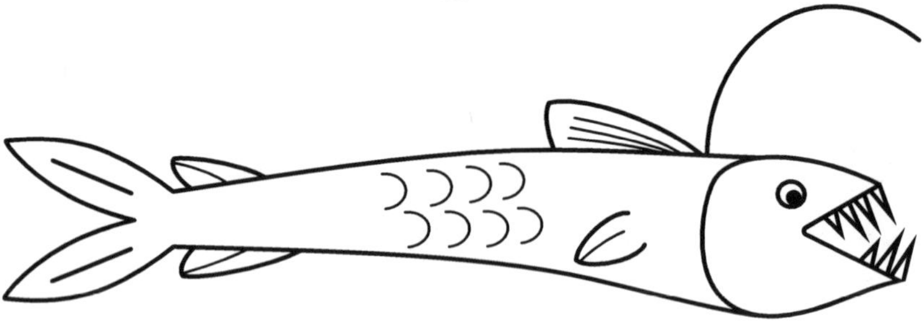

Viperfish

Ww

Walrus

Xx

X-ray Tetra

Y y

Yellow Tang

Zz

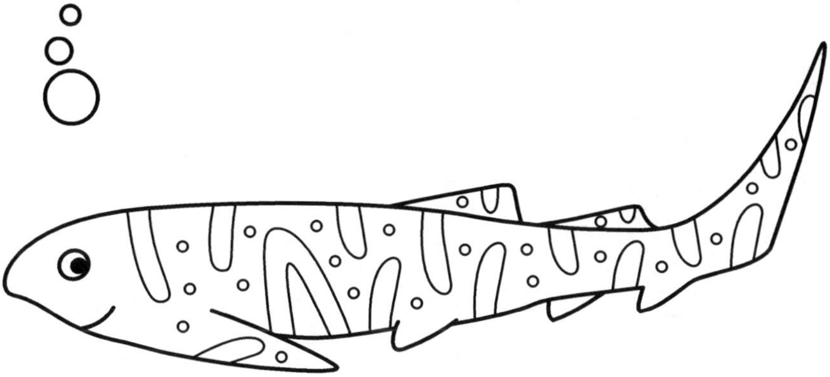

Zebra Shark
(baby)

Numbers and Vision Tracker

Section 3

From 0 to 9,
count them all

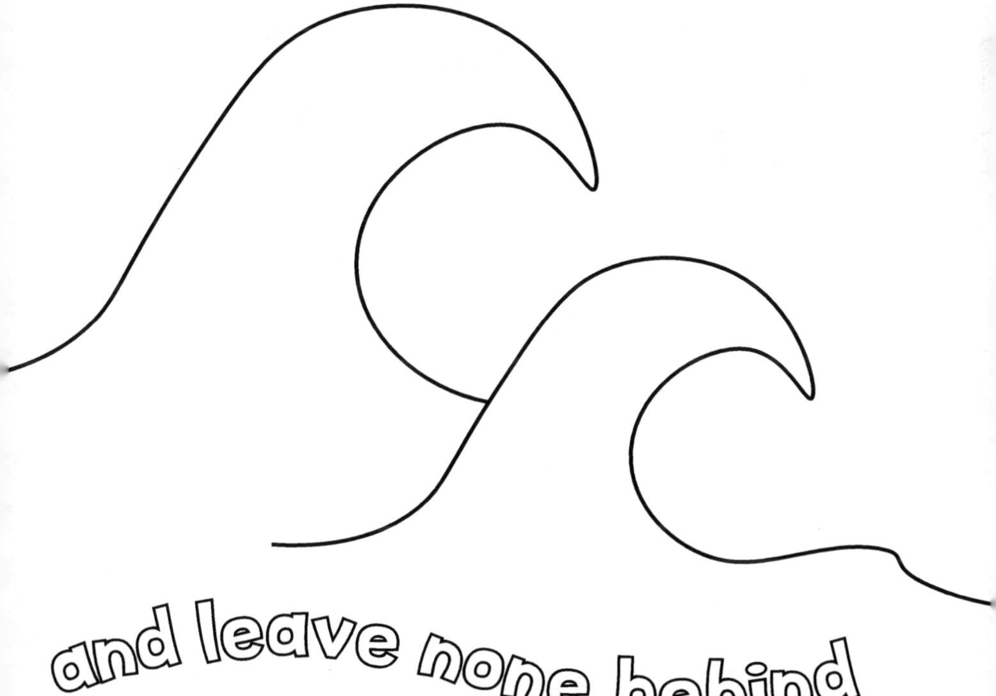

and leave none behind

0

Zero

1

One

2

Two

3

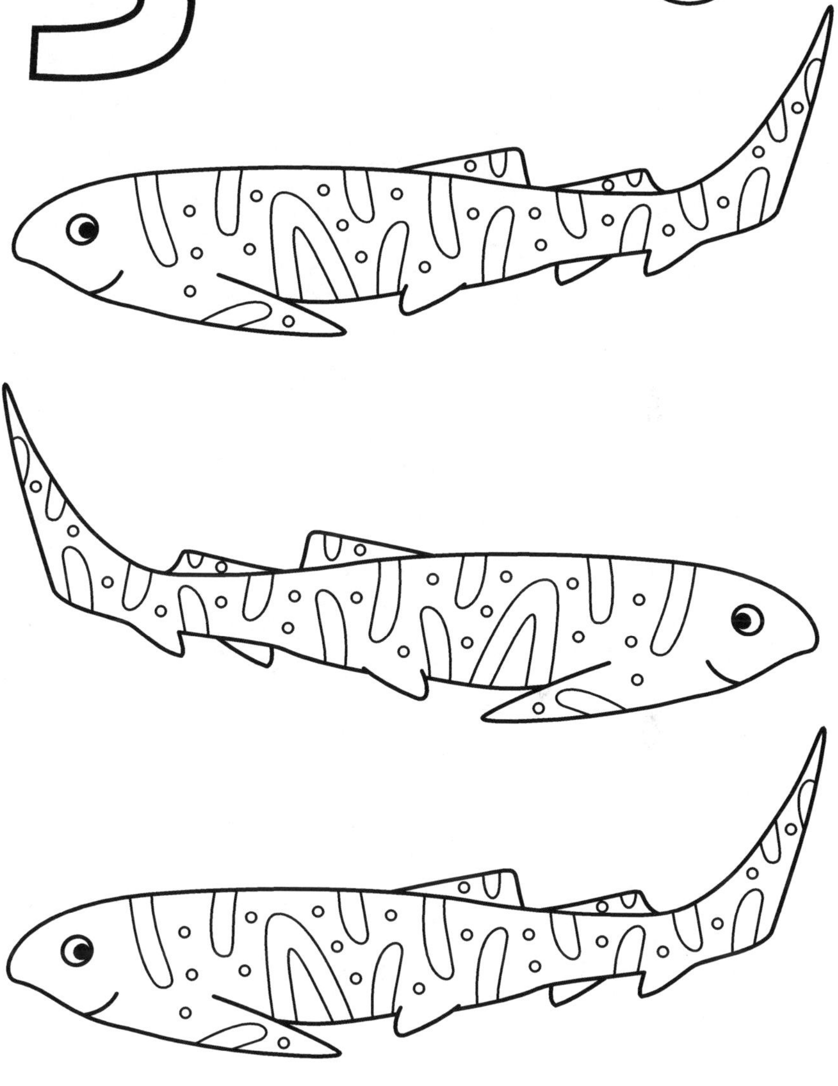

Three

4

Four

5

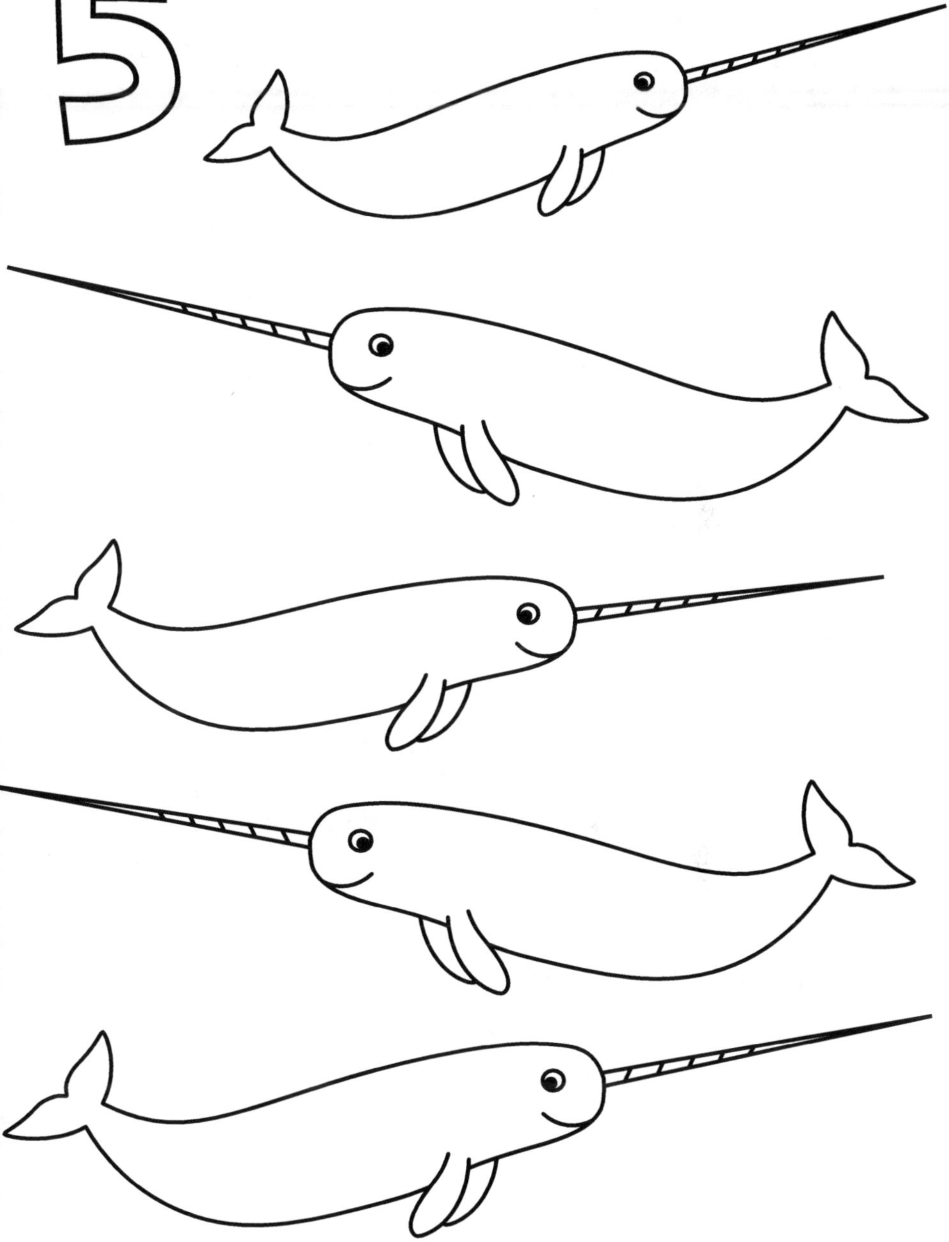

Five

6

Six

7

Seven

8

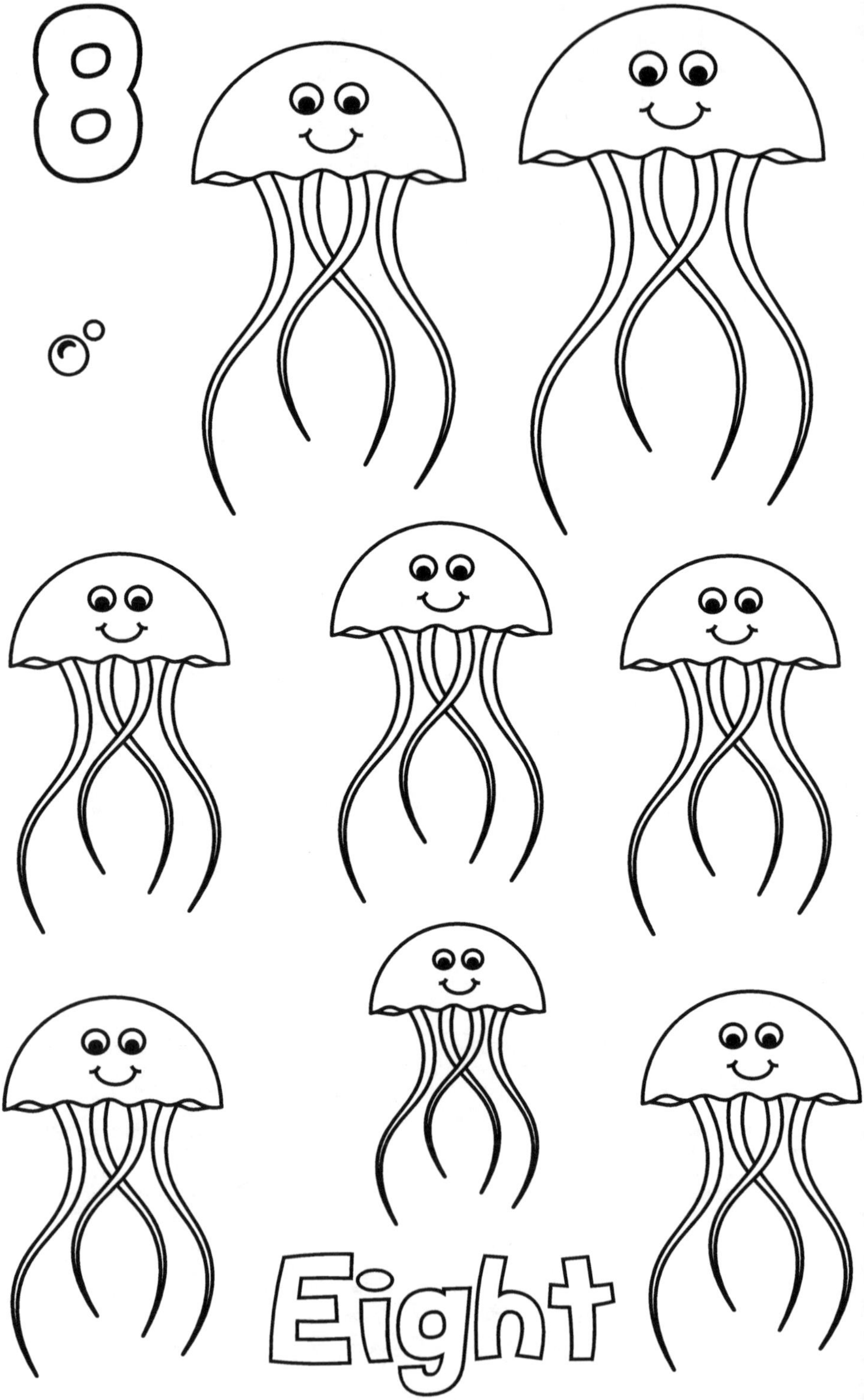

Eight

9

Nine

Amblyopia (lazy eye) refers to reduced vision in one eye which develops early on in life. It can also occur in both eyes but this is rare.
Quite often kids require glasses before patching treatment for amblyopia is started.

Cover design, story, illustrator & publisher: V.Heeralal

Email: quicktips.oporthics@gmail.com

Website: oporthix.com

ISBN: 978-0-473-60285-7

Published in New Zealand

Oporthics: *A combination of the words 'Optometry' and 'Orthoptics'.*

**Combined by the illustrator specifically for the email address*

Made in the USA
Monee, IL
04 November 2026

67375785R10026